"After lunch we will hunt," said Miss Lerner.

"We will hunt a bug, a bird, and a plant."

"First, we do not pick a plant.
Number two, we do not pick up a bug.
We just spot it."

"I spot an ant," said Darrell.

"A bird!" said Amber.

"What is under this?" said Kirk.

"This plant is a fern," said Miss Lerner.

The class sent her a letter.